How to build a digital microscope

- construct a reliable, inexpensive, microscope
for both regular and polarized light microscopy

Lasse Lu Pedersen

DEDICATED TO

Rong and Louis

CONTENTS

INTRODUCTION

The microscope is the most important tool in the microbiologists toolbox. Nowadays, several companies sell very well made microscopes, but such a luxury was unknown to the earliest microbiologists. They had to build their own microscopes, and many advancements to the science of microbiology were made by people who managed to improve on their own private microscopes.

While basic models have become increasingly affordable in recent years, purchasing a digital microscope can still be a substantial investment - and one that many interested individuals, families, and schools are unwilling or unable to make.

Fortunately, it has also become increasingly easy to get the parts for building your very own microscope at a price that is more approachable than that of commercial microscopes!

How cheaply you can build a digital microscope will depend on what materials you have readily available and on which design you choose to go for - but the core optical and digital components of a   digital microscope can be had for as little as $4.50 at eBay. This covers the cost of one microscope lens, one digital camera with USB connection and a light source suitable for building a bright-field microscope, which is a microscope in which the light from the light source passes though the sample being studied and into first the lens and then further on into the camera. The $4.50 even covers shipping from China to Europe or the States. Add $1.00 more and you can have one set of filters capable of turning your microscope into a polarized light microscope - a microscope capable of letting you see rocks, crystals, and other things in a whole new light! To give you a point of reference: the cheapest, used, digital polarized light microscope I can find at eBay today would set you back $2250 plus shipping.

The cheap homebuilt microscope will not have the specifications of the expensive factory made one, but it will can be extremely good value for money, it will let you explore microscopy at a budget, and building it can be an entertaining and educational project for a family or a young aspiring scientist.

Here I will present two homebuilt digital microscopes, you can choose to copy one of them, or you can use them as inspiration to build your own microscope. They offer x100 and x250 magnification, can be made to connect to a computer using USB, or you can opt for using a smartphone to view and record the images - and both can be built as a regular optical microscope or as a polarized light microscope.

Once you have read this book, have build yourself a microscope and have had a bit of time to play with your new tool, sit down and ponder your achievement.

Consider if it has been worth your time.

And if you are able to sponsor one more microscope.

If you are inclined to reply 'yes' to these two questions, then think if you know a kid who would enjoy building a microscope with you and would have a good time investigating the world with his or her very own microscope.

Thank you!

BE RESPONSIBLE

If you are underage then please ask an adult to supervise the construction of the microscope to ensure that all tools and materials are being used correctly and safely.

In return for the favor you can let them borrow the microscope now and then - or *you* can help *them* build their own microscope!

HISTORY OF OPTICAL MICROSCOPY

The first lenses were made by ancient Egyptians and Mesopotamians who manufactured them from polished crystal as early as 750 BC. Much later, in the Middle Ages, the first lenses made from glass saw the light of day, and in recent times cheap, inexpensive lenses are being made from various plastics and incorporated in many consumer goods such as laser pointers and optical disc drives.

We don't know who was the first person to put several lenses together to create the first compound microscope, a microscope using several lenses to achieve a high degree of magnification. Some sources claim that it was the Dutch spectacle-maker Zacharias Janssen who invented it in 1590, others that it was his competitor Hans Lippershey.

What we do know is that combining several lenses into one apparatus allows for much higher magnification than simple lenses. The word 'microscope' was introduced by Giovanni Faber, a German doctor and botanist, who used it to describe the famed astronomer Galileo Galilei's compound microscope in 1625.

Galileo himself had referred to his instrument as the 'little eye', the 'occhiolino'.

The first account of a microscope being used to study living tissue appeared two decades later when Giovanni Hodierna published his studies of insect eyes in his book 'The Fly's Eye'. In the next decades microscopes were used to study many different biological samples, and in 1676 Antonie van Leeuwenhoek reported that he had discovered microorganisms, a feat that has earned him recognition as the world's first microbiologist. Thanks to the powerful lenses he crafted he also made other significant contributions to microscopy, such as discovering red blood cells and describing how they flow in capillaries, the smallest blood vessels in the human body.

At the end of the 17th century Antonie van Leeuwenhoek was responsible for most major discoveries made using a microscope. While he was undoubtedly a great craftsman and researcher, part of the reason why he had such a monopoly was that he refused to show others his best lenses, and even tried to mislead people by showing visitors lenses of inferior quality and not sharing his lens making technique with anyone. It took more than a century before compound microscopes were able to match the image quality that one of his lenses could provide.

Times have changed. In recent years lenses have become increasingly accessible and affordable, and the same has happened to digital cameras, which means that today you can build your very own digital microscope, take pictures of what you study and of your microscope - and then share them with everyone!

LIGHT MICROSCOPY

Most simple lenses have two surfaces which can be either planar, bulging outwards from the center (convex) or depressed into the lens (concave). They fall in two categories, converging lenses which will focus a beam of light into a point, and diverging lenses which spread light passing through them.

Antonie van Leeuwenhoek made his lenses by heating the middle of a thin glass rod in a hot flame and subsequently pulling the rod apart. This created two long, thin glass threads, each of which could be placed in the flame again, creating a small, high quality glass sphere. His lenses were converging lenses, and the smaller the sphere, the higher magnification it provided.

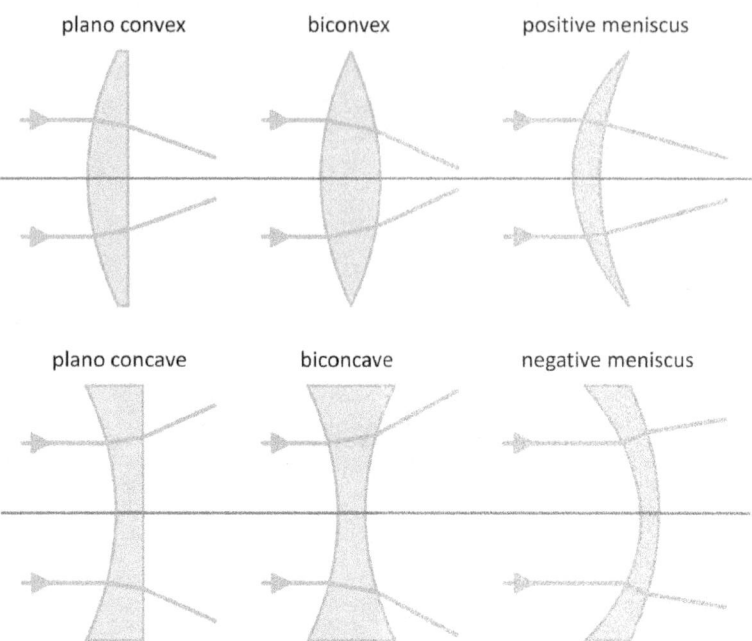

The point in which a converging lens focuses light is the lens' **focal point**, the distance from the lens to its focal point is its **focal length**, the line going though the centers of its two surfaces is its **optical axis**, and the plane going though the focal point perpendicular to the optical axis is its **focal plane**.

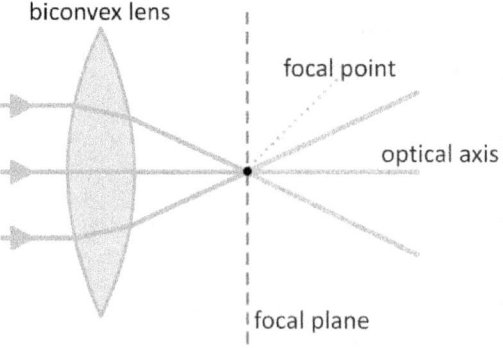

When a beam of light is spread by a diverging lens it appears to spring from a point on the optical axis in front of the lens. Here the point is also called the focal point, and the distance from it to the lens is likewise the focal length, but compared to the focal length of a converging lens this one is negative.

A digital bright-field microscope is composed of four core components: at the bottom sits the **light source** which sends light though the specimen (the sample being investigated). The specimen has been placed on a microscope slide which is held by the **focusing stage**, which can be moved up and down. This allows the user to place the specimen in the focal plane of the lens, which is known as 'focusing on the specimen'. Then comes a **lens** which collects the light that passes though the specimen, magnifies the image of the specimen and sends it to the eyepiece, or in the case of a digital microscope, the **camera**, where it can be viewed.

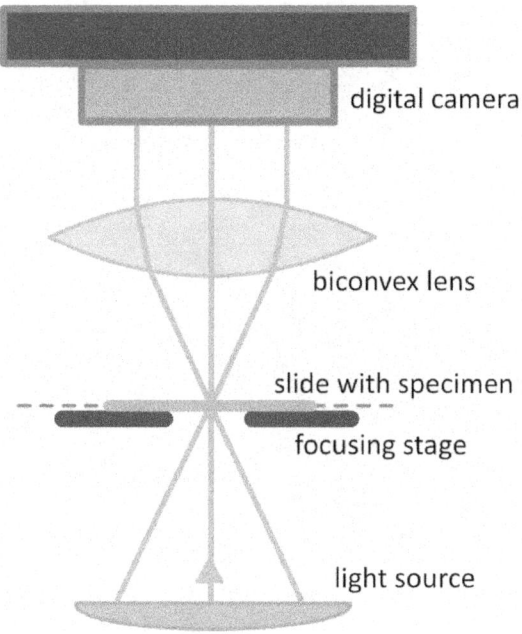

digital camera

biconvex lens

slide with specimen

focusing stage

light source

POLARIZED LIGHT

A beam of light is composed of many light waves (think of them as 'units of light') which travel in the same direction. Each light wave has its own orientation, designated its 'polarization'. For example: one light wave might be thought of as "swinging" from up to down and then up again while moving forwards, while another wave is rotated relative to the first wave, but still moving in the same direction. The two light waves are part of the same beam of light, but they have different polarizations, and the beam of light is said to be **'unpolarized'**.

If a beam of unpolarized light is passed though a special filter known as a linear **polarizer** which only lets though light waves with one specific polarization, then the light beam becomes **'polarized'**. If the polarized beam is then passed though a second polarizing filter which has been rotated 90° relative to the first filter, then no light should pass though the second filter unless something in between the two filters has rotated the beam. Since the second filter tells if such a rotation has taken place it is named the **'analyzer'**.

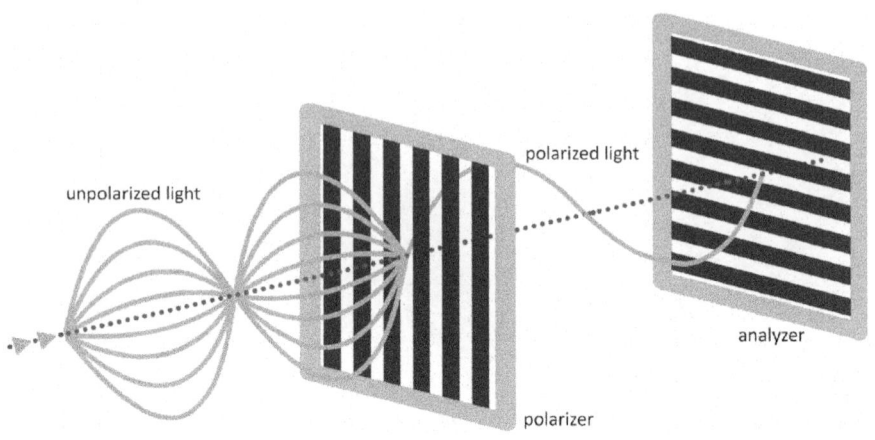

Polarizing filters can be incorporated into a bright-field microscope, which will allow the user of the microscope to see if the specimen, or parts of it, is able to rotate light.

The polarizer is placed before the specimen and the analyzer is placed just before the image reaches the camera. If anything between the polarizer and the analyzer has rotated the light, then it will be seen as colorful areas on the specimen.

Air, water, glass, and acrylic glass do not rotate light, so all the rotation that takes place between the polarizer and the analyzer can be described to the specimen being examined. Materials which rotate light include many minerals and crystals, and polarized light microscopes are often referred to as 'petrographic microscopes' (petrology is the study of the origin, distribution, composition and structure of minerals).

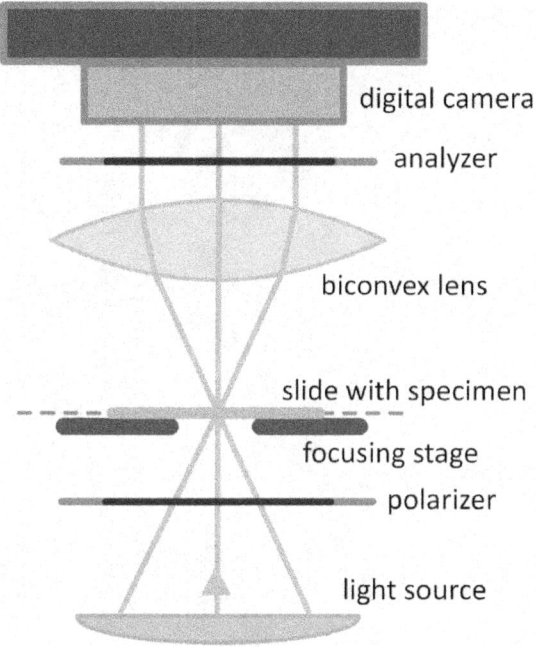

digital camera

analyzer

biconvex lens

slide with specimen

focusing stage

polarizer

light source

Professional petrographic microscopes come equipped with a circular rotating focusing stage instead of the regular focusing stage, and has two additional optical parts called a waveplate and a Bertrand lens, which help in analyzing the specimen.

Building a circular rotating focusing stage which has the same functionality as the commercial ones takes precision machining which is beyond the scope of this project, and neither a waveplate nor or a Bertrand lens can currently be acquired at a low budget. Fortunately the basic digital polarized light microscope presented here enables its user to examine crystal and mineral structures and record beautiful images and videos without these parts.

A crystal photographed in unpolarized (top) and polarized (bottom)

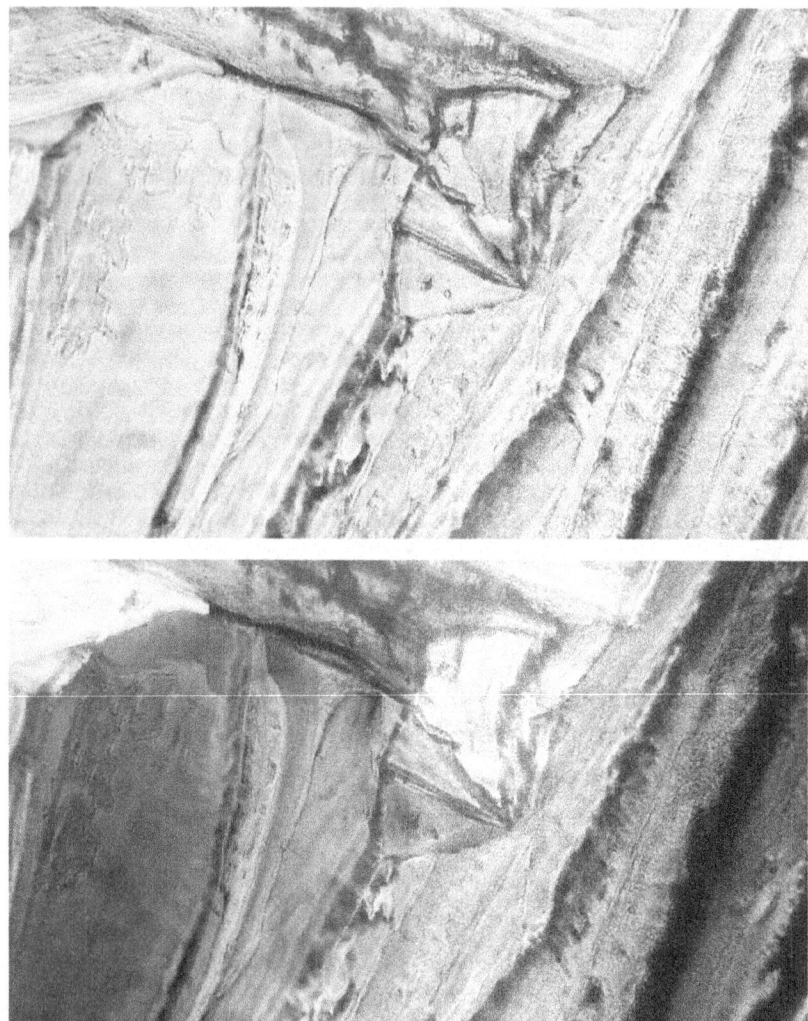

light, at x250 magnification.

MATERIALS NEEDED

As already mentioned a digital microscope essentially consists of a light source to illuminate the object you want to study, a lens to magnify its image, a digital camera to record the magnified image and a focusing stage which allows you to focus on the object you are studying... and then you need something to hold all these parts together.

The light source

I suggest that you get your hands on a mini LED flashlight at a local dollar store or from eBay, it should set you back roughly $1. If you have good knowledge of electricity, electrical circuits and know how to solder then you can also build your light source from scratch , but that is purely optional, and is primarily for people who have very specific desires when it comes to their light source.

The lens

When it comes to scavenging for a lens chances are that you will end up with a plano convex or possibly a biconvex lens made from acrylic or some other kind of plastic. And that is great, as many of these will work perfectly in your microscope! I suggest you look for one that has a diameter of roughly 7mm (½"). That said, most people can't just go and buy just a lens in a local shop, so there is probably a limited to how picky you can allow yourself to be.

You may be able to find a lens by searching for 'focusing lens for laser diode' at eBay, one lens should set you back roughly $1, but chances are that you will want to opt for extracting one from a cheap laser pointer, the kind that you can get for about $1 at a dollar store or eBay. If you are slightly more adventurous you should be able to get one from disassembling an old CD, DVD, or Blue-Ray drive. I've successfully built microscopes using lenses from all these sources which have worked perfectly, but there is a risk that you won't get a good lens from your first laser pointer or drive.

Using just one of these lenses should give you roughly x100 magnification, and two can be combined to give in the vicinity of x250 magnification, so consider buying two right away. Sadly using more than two lenses tends to give very poor images, so x250 magnification is about as much as you can expect from these lenses.

The digital camera

Many smartphones come equipped with a nice digital camera which will work splendidly for our purposes, but this option hardly qualifies as 'inexpensive ', so I have a good alternative for you: a $2.50 webcamera from eBay. You should look for a model which has a lens no bigger than 5mm (5/12"), and which can lie stably with the lens facing downwards.

The focusing stage

While the lens, the camera and the light source may seem to be the most important components of the microscope, the focusing stage is the part that is most likely to cause you headaches, literally. Without a good focusing stage it will be difficult to focus on your specimen, and that will make it difficult to get sharp images from your camera.

The first microscope design utilizes a focusing stage composed of a mini LED flashlight, a 1" metal double compression fitting and a 1" female-to-female threaded pipe fitting - the latter two can be had at eBay for $4.99 and $2.49, respectively. This is more than all the other components combined, but it will give you a reliable focusing stage with little effort.

The second microscope design has a focusing stage build from acrylic glass, two bolts, two nuts and some washers. This approach can also give you a good focusing stage, but getting it all aligned just perfectly may take a few tries.

Holding everything together

I prefer to build the microscope frame from acrylic glass or a combination of acrylic glass, nuts and bolts. While acrylic glass isn't cheap, many shops selling it will let you have some of their offcuts either cheap or for free if you ask nicely. I recommend that you use acrylic glass that is roughly 1mm thicker than your lens(es). You will have to measure your own lens to be certain, but most lenses from laser pointers and optical drives are around3mm thick. So if you are using one lens, then use 4mm thick acrylic, and if you are using two lenses, then use 7mm thick acrylic. You can use thicker acrylic, but due to the focal length of the lenses mentioned here you should avoid using sheets thicker than 8mm.

If you aren't using nuts and bolts to hold the acrylic sheets together you will also need some glue suitable for acrylic glass. I use Acrifix® 1R 0192, but instead of trying to get this specific glue I suggest you ask the shop where you got your acrylic glass. Superglue works fine for bonding acrylic - though some variants may give the joints a frosted look.

If you can't source acrylic glass then you can use other kinds of plastic or even wood. Do note that if you don't use a clear material you will need to drill an extra hole in the focusing stage, to make sure that the light beam coming from the light source isn't blocked. A 10 to 15 mm hole in the focusing stage right under the spot where the specimen is when it is in focus should suffice.

Filters for converting the microscope to a polarized light microscope

In order to convert a classical microscope to a polarized light microscope you will need to get your hands on two linear polarizing filters. The cheapest option is to buy a pair of linear polarizing 3D glasses ($0.99 at eBay). Alternatively you can opt for linear polarizing filters meant for SLR cameras, or you can buy a two pack of linear polarizing laminated film from edmundoptics.com for $12 (stock number 43-781).

Note that there is also something called a circular polarizer filters, do not buy these.

HOW TO BUILD A
SIMPLE DIGITAL MICROSCOPE

Materials used

Four acrylic glass sheets, 110 x 130mm, 7mm thick

One lens from a laser pointer

One mini led flashlight

One 1" metal double compression fitting

One 1" female-to-female threaded pipe fitting

One webcamera / smartphone

Glue suitable for acrylic glass

Assembly instructions

Take one of the four acrylic glass sheets , this sheet will be the one that contains the lens(es). The hole for the lens should be roughly 30mm from the shortest side of the sheet.

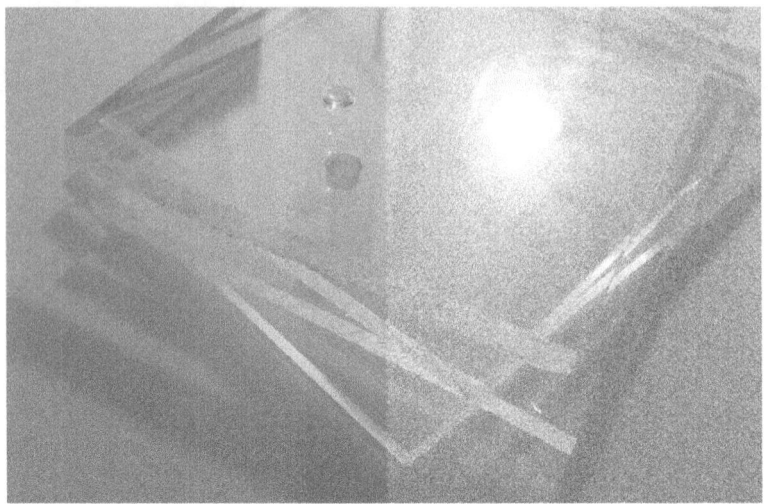

Do note that drills make holes with a diameter slightly larger than their own diameter, and it is difficult to reduce the diameter of a hole once it has been made. Consequently one has to take care when making the hole for the lens. If you drill the hole too big your lens will fall though it, and you'll have to drill a new hole and explain to your friends why your microscope has that extra hole.

One way to drill a good hole for the lens is to first drill though the sheet with a drill a bit smaller than the lens, and then drill in the same hole with a drill slightly bigger than the lens, but only drill though two thirds of the thickness of the sheet. Another way of accomplishing the task is to drill the hole with a drill smaller than the lens, and then expand the hole using a round file so that the top of the hole is slightly larger than the lens, and the bottom is slightly smaller. The former way is shown first, and the latter below it.

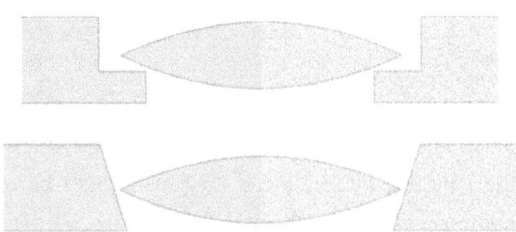

If you want to be able to use both x100 and x250 magnification then you can use two lenses in the same hole, and remove the top one when you want to use x100 magnification. Alternatively you can have two holes, one with two lenses in it, and another one with one lens in it.

Once the hole has been made you should glue the four sheets together as shown below, and then insert the lens(es) into the hole.

Now to make the combined focusing stage and light source.

Take apart the 1" double compression fitting and keep the male-to-male threaded pipe fitting and the two compression nuts from it. Then glue the 1" female-to-female threaded pipe fitting to the top of the flashlight using superglue. Try to avoid getting glue on the middle of the flashlight lens, as this will make part of the lens opaque and reduce the amount of light that reaches the specimen.

Wait until the glue on the female-to-female fitting and the flashlight has dried completely. Screw the male-to-male fitting into the female-to-female fitting, and then screw one of the compression nuts onto the free end of the male-to-male fitting. Glue the other compression nut to the bottom of the flashlight as shown below. If the flashlights on-off switch is located on its bottom then avoid getting glue into the switch. The reason for adding the compression nut to the bottom is that it makes the focusing stage less top-heavy.

Now place the combined focusing stage and light source under the lens and place a microscope slide on top of the stage.

Now the specimen on the slide can be inspected by placing the phone on top of the microscope and focusing by screwing or unscrewing the male-to-male fitting and the upper compression nut. If the focusing stage isn't tall enough to place the specimen in the lens' focal point, then you can add a few washers under the bottom compression nut.

Note that it is easiest to align the phone and the microscope if you place the phone on the microscope while the camera is recording.

Bamboo cane, x250 magnification

If you are using a webcam then the procedure is the same as for the phone.

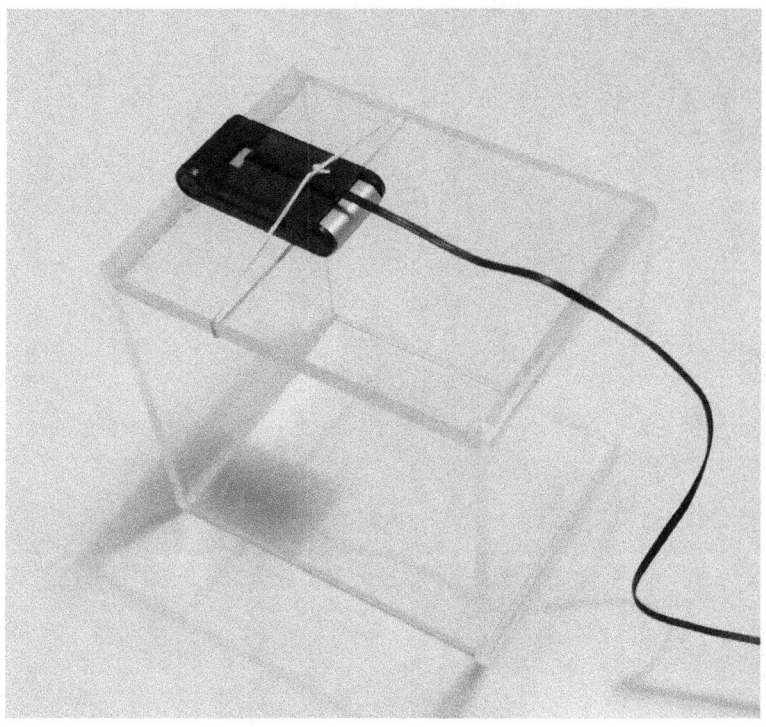

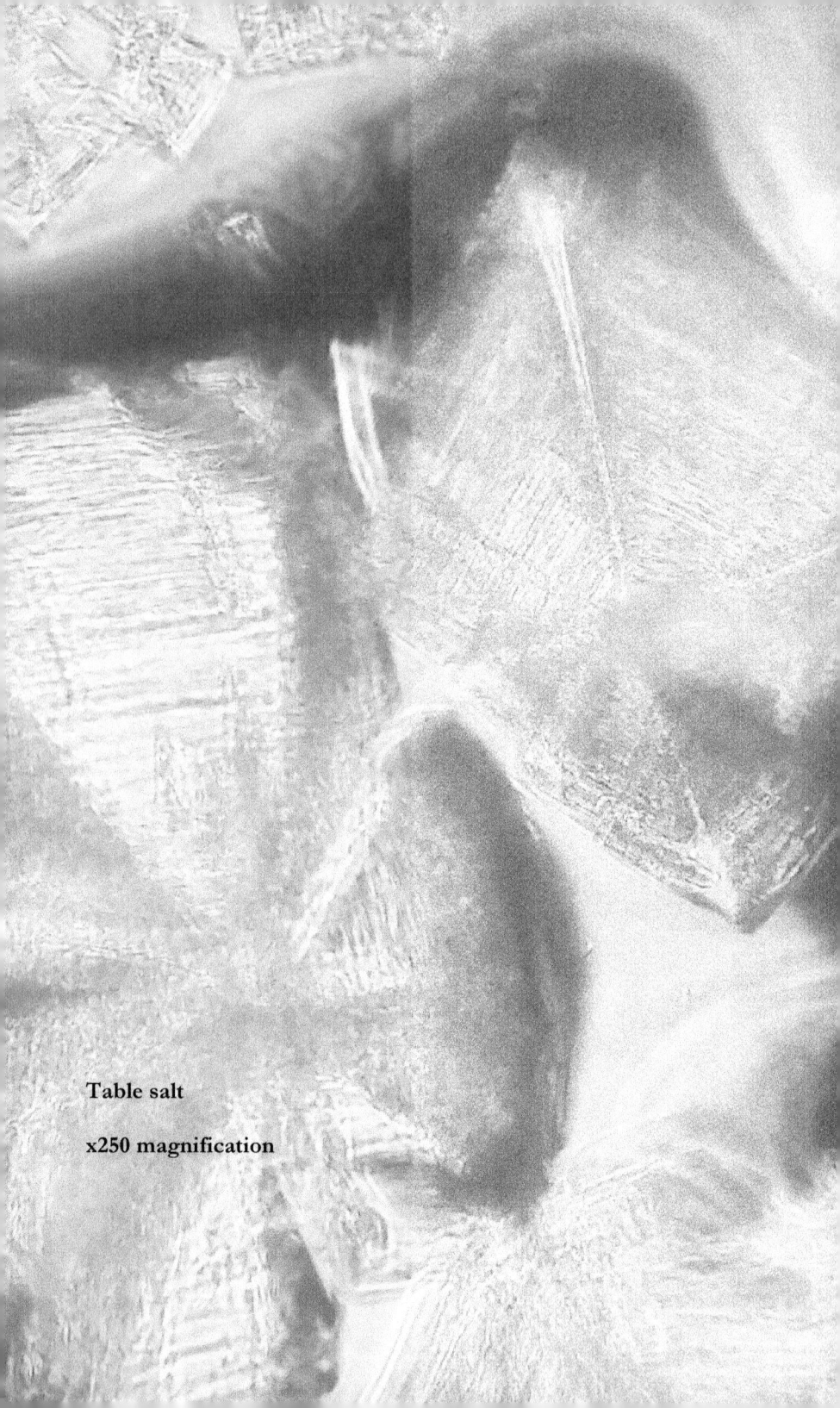

Table salt

x250 magnification

CONVERT A
REGULAR LIGHT MICROSCOPE
TO A POLARIZED LIGHT MICROSCOPE

This method should in theory work for all bright field microscopes, both home build and factory made ones. In some cases the manufacturer might have made the conversion difficult or practically impossible, so make sure to consult them before attempting a conversion, to make sure you don't ruin the microscope.

Materials used

Two linear polarizer filters

Assembly instructions

Simply place the first linear polarizer filter (the polarizer) between the light source and the microscope slide, the second filter (the analyzer) between the lens and the camera/phone and turn one of the filters clockwise or counterclockwise until the filters together block all light (or as close as you can get, it is unlikely that cheap filters will block 100%) from the light source that doesn't pass though the slide. Ideally one of the filters should be positioned such that it is possible to turn it while a specimen is being studied.

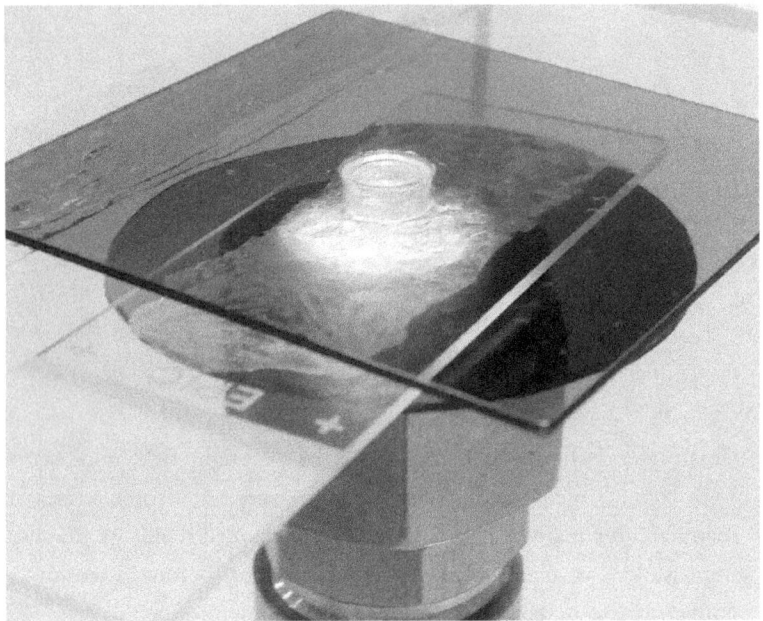

Sucrose crystal in polarized light

x100 magnification

HOW TO BUILD A DIGITAL POLARIZIED LIGHT MICROSCOPE

The second build is more complicated than the first, it utilizes a custom light source that lets you switch between white and UV light and adjust the intensity of the two types of light independently. It has a different focusing stage, which doesn't require a compression fitting but can be build from inexpensive acrylic glass offcuts, bolts, nuts and washers.

If you don't have experience with electric circuits and soldering and *know with certainty* that you can build the light source presented here safely, then either have an experienced person build this part for you, or use a readymade one such as a LED flashlight, as shown in the previous build.

Materials used

One acrylic glass sheet, 130 x 200mm, 7mm thick

One acrylic glass sheet, 120 x 55mm, 7mm thick

One plastic project box, 120 x 190 x 60mm

Two lenses from DVD drives

Two linear polarizer filters

One webcamera / smartphone

Three neutral white LEDs - 690 lm @ 700mA

Three UV LEDs - 900 lm @ 700mA

Two LuxDrive 700mA A011 FlexBlock Constant Current Drivers

Two 20k ohm linear potentiometers

Two SPST switches

One aluminum heat sink, 50 x 50 x 35mm

Two 12V computer fans, 40 x 40 x 10mm

One 12V 5A DC power supply

Solid core wire, heat shrink tubing, and solder

Assorted nuts, bolts, and washers

Rubber band

Arctic Silver thermal adhesive glue

Superglue

Assembly instructions

This time I'll incorporate one of linear polarizer filters right away - the second one can be placed on top of the lens whenever one wants to use the microscope for polarized light microscopy. It is true that this will make the light source less powerful when used for regular light microscopy, but the LEDs I'm using are so powerful that that isn't of concern. The filters I'm using are two used SLR camera linear polarizer filters.

The light source will consist of six LEDs, three white and three UV. The three white light LEDs come on one small heat sink, while the UV LEDs have separate heat sinks.

Do note that UV light can damage your eyes and skin, *do not* use it without proper protective equipment. I have added UV light primarily to illustrate how you can incorporate two light sources into your microscope. This may be useful if you want to study specimens you know absorb light with a specific wavelength.

Here we have the two types of LEDs used in this build, three LuxeonStar neutral white LEDs on the left, and one LedEngin LZ1-00UA00 UV led on the right. If you want to build this microscope then I suggest that you look for LEDs with similar performance characteristics, instead of trying to get exactly the same LEDs, as that increases the likelihood of a local supplier having a good offer for you.

The LEDs I use generate a lot of heat, and require cooling in the form of a heat sink and fans.

Power is supplied by a 12V 5A DC power supply, and the controls are made up of constant current LED drivers, linear potentiometers and SPST switches.

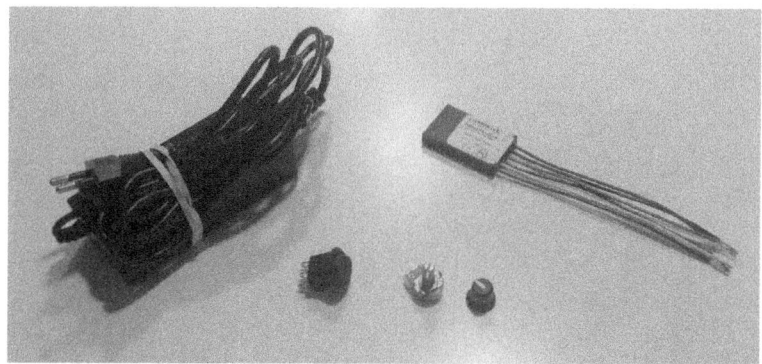

All the electrical components are connected using electrical wiring, insulated with heat shrink tubing and contained in a plastic project box.

First I cut a hole for one of the SLR camera polarizing light filters in what will be the top of the light source enclosure. Once the filter has been fitted into the hole it can be turned to change the angle of the polarized light.

Holes are cut for the switches, potentiometers, power supply, and fans.

And everything is soldered together as per this diagram.

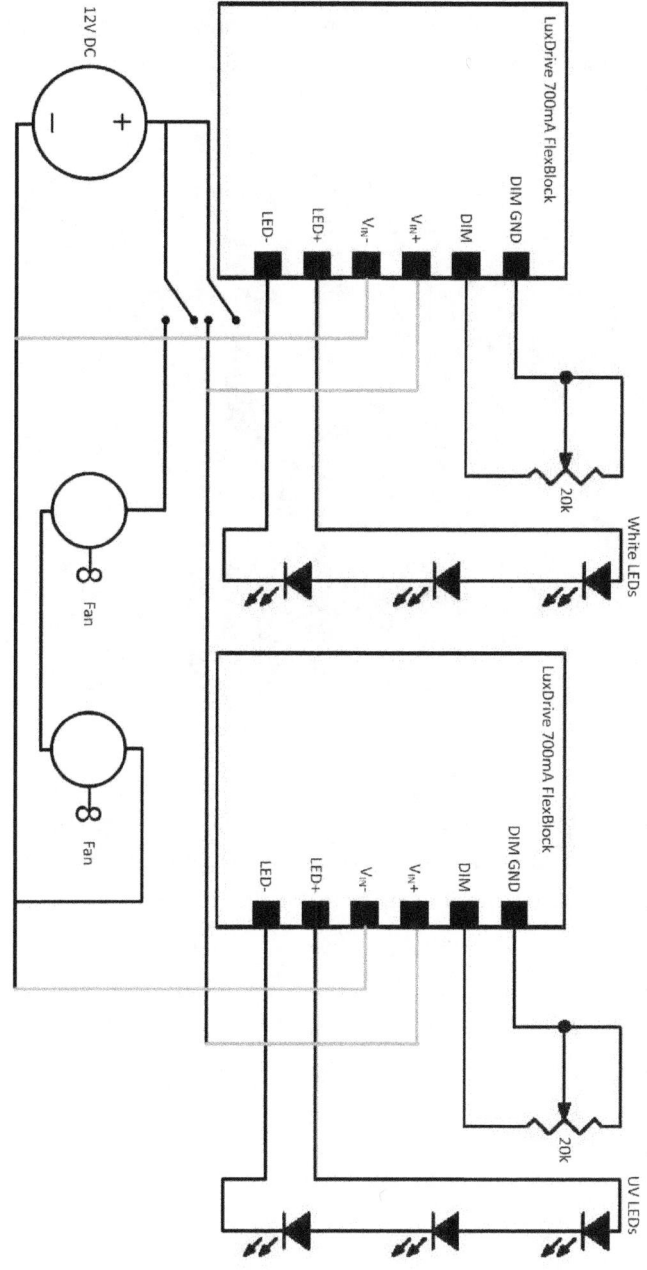

It is easiest to solder the LEDs before you mount them on the big heat sink using the thermal adhesive glue. If you mount them on the heat sink before soldering then the heat sink will draw the heat from the soldering iron away from the contact areas, making it difficult to get a good bond.

Everything is fitted into the project box. The light source, fans and constant current LED drivers are glued in place using superglue - avoid getting glue on the moving parts of the fans, as this will ruin them.

First the white light is tested.

And then the UV light is tested.

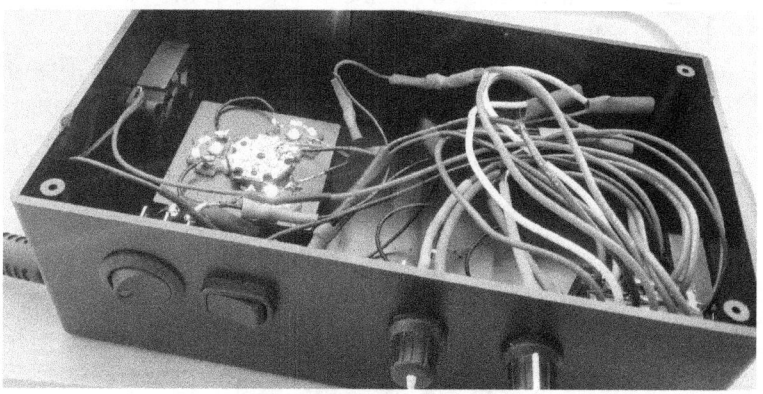

And this is how the light source looks with the top of the enclosure in place, but without the rest of the microscope attached.

Next step is to attach the 130 x 200mm plate which will hold the lenses and the focusing stage. Here I have drilled holes in the corners of the project box, superimposed them on the acrylic plate and drilled four matching holes in it, and then connected the two pieces of plastic using M8 bolts and nuts. While the plate is attached I find the point on the acrylic sheet which is directly above the center of the polarizing filter, and mark it with a pen.

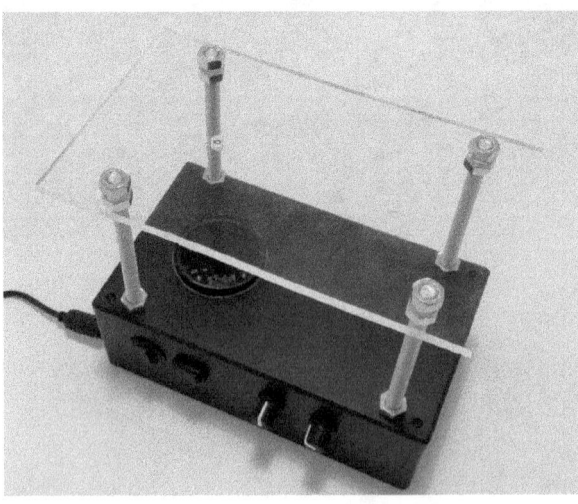

Now take the 120 x 55mm acrylic sheet and drill two holes as shown below, roughly 10mm from the short sides. Superimpose these holes onto the underside of the big acrylic sheet such that the small sheet won't hit any of the corner bolts, and the middle of the small sheet is right under the spot where the hole for the lens will be. These holes are for holding the focusing stage in place.

Now remove the large acrylic sheet from the four bolts and drill the holes for the lenses and for holding the focusing stage. The latter should not go all the way through the sheet, don't drill them deeper than two thirds of the thickness of the acrylic sheet. This ensures that the top surface of the large acrylic sheet (where your camera or phone will lie) is smooth.

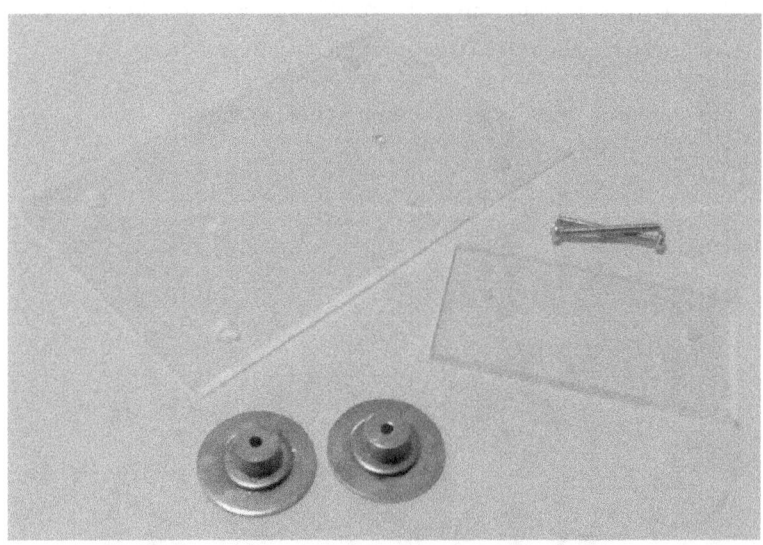

For adjusting the focusing stage I have manufactured two knobs from washers, two M4 nuts (one nut per knob) and a few drops of superglue. The big washers supports the small acrylic sheet, and the small washers will be connected using a rubber band.

M4 bolts are screwed into the knobs, passed though the small acrylic sheet and then gently screwed into the underside of the large acrylic sheet, which is then fixed on top of the four large bolts.

The lenses are inserted into the central hole in the large sheet.

Right now the microscope is quite fragile, due to the way the focusing stage is attached to the rest of the instrument. Now is the time to perform a short test: place a microscope slide on the focusing stage and your camera or phone on top of the large acrylic sheet, with the camera looking into the lenses inserted into the sheet. Carefully focus on the slide using the knobs.

If you are able to focus on the slide and the focusing stage goes up and down smoothly, then gently remove the camera / phone and the slide, and apply a small drop of superglue where the two bolts from the focusing stage enter the large acrylic sheet. Avoid having the superglue run down the bolts towards the focusing stage, as this will ruin the threading on the bolts.

If you can't focus on the slide you have to adjust the focusing stage until it works properly.

Once the superglue has dried you can connect the two knobs using the rubber band; it should be so tight that turning one knob results in the other knob turning at the same time, but not much tighter than that, as that puts unnecessary strain on the focusing stage.

Now the microscope is ready for use!

If you want to use it for polarized light microscopy then place the second linear polarizer filter between the camera/phone and the lens (if your linear polarizer filter is the SLR camera variant then you will have to remove the second filter from the black adapter rings), if you want to use it as a classical microscope then simply leave the second filter out.

Here the microscope is used to examine a slide carrying a thin section of marble using polarized light.

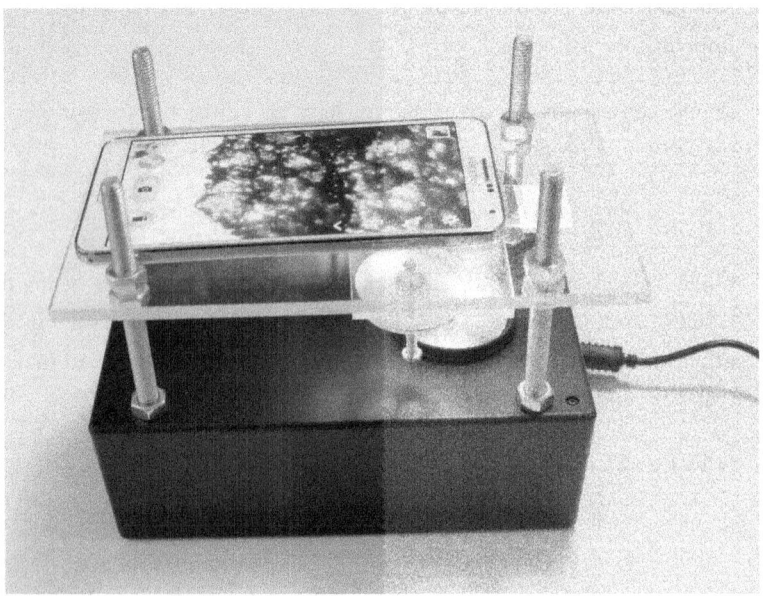

Thin section of marble in polarized light

250 magnification

Lasse Lu Pedersen

YOUR VERY OWN MICROSCOPE

Now you've seen two different home build microscopes, and it is time for you to decide how you want to build your own! If you are looking to build your microscope on the lowest possible budget, go for the first design presented in this book, but replace the focusing stage with the one used in the second design.

Start by finding a shop selling acrylic glass and politely ask if you can have some of the offcuts they won't be using anyway. If they also have some small, thin pieces in addition to the ones you need for the microscope itself, then you can cut them into improvised microscope slides (more on slides in the next chapter).

Then get a laser pointer (or two, if you want x250 magnification) and a mini LED flashlight from a dollar store, and a cheap webcamera from a local computer store or eBay, if the local store isn't cheap enough. If a local cinema shows 3D movies then you might be able to source a pair of linear polarizing 3D glasses there, if not then it is off to eBay again. That leaves a few nuts, bolts, washers, a rubber band and a few drops of superglue, which can be had at most hardware stores.

Remember that if you can't get acrylic glass or another clear material you will need to drill an extra hole in the focusing stage, to make sure that the light beam from the light source isn't blocked. A 10 to 15 mm hole right under the spot where the specimen is when it is in focus should suffice.

MICROSCOPE SLIDES

The function of a microscope slide is to hold an object, usually referred to as the 'specimen', for examination under a microscope. They are made of glass or plastic, usually measuring 25 x 75 mm (1 x 3") and 1 to 1.2 mm thick. You should be able to get 50 glass slides for approximately $6 at eBay, but if your budget doesn't allow for this you can use any ~1mm thick clear plastic sheet that you cut to size.

Microscope slides are regularly used together with cover slips (you usually get 100 for $2 at eBay), very thin square pieces of glass that are placed on top of the sample on the microscope slide. Be very gentle when handling them, they are fragile and break easily. Their function is to flatten out the specimen, making it easier to focus on it. Additionally, they also serve to protect the lens from whatever specimen is being studied.

The most common way of preparing a microscope slide is the 'wet mount' slide, in which you place a drop of sample (such as water from a pond or small stream) on a slide and then lower a cover slip gently over the drop at an angle. Do not apply pressure to the cover slip, merely let the liquid spread between the two layers of glass. You may need to practice how much liquid to use for a wet mount slide: if you use too little your slide will dry up too fast, and you risk crushing organisms in the sample, and if you use too much the cover slip will slide along the microscope slide.

A slide prepared using just the right amount of liquid can be expected to last up to 30 minutes before it dries out due to heat from the microscopes light source and from the surroundings. If you need more time than this then you can apply petroleum jelly onto the four edges of the cover slip before you place it on top of the sample (jelly facing downwards). Try to avoid getting air bubbles caught under the cover slip. Here you will need to press lightly on the cover glass to ensure you get a good seal. If done correctly the slide may last two days or more. Microscope slides and cover slips used in wet mounts can be rinsed and reused many times.

Another type of microscope slide is the depression slide, which has a small well in the center. It is useful when you want to observe animals 1-5 mm in length such as Daphnia (a group of small aquatic crustaceans found in some streams, rivers, ponds and lakes) which would be crushed under the cover slip in a wet mount slide. To use a depression slide transfer a drop of sample containing the Daphnia to the well and touch one edge of the drop with a bit of paper towel. Soak up just enough water to make the Daphnia unable to move, which will make it easier to observe using the microscope. Depression slides are significantly more expensive than flat slides.

If you want to buy prepared microscope slides then $12 or so should buy you a set of 48 prepared microscope slides at eBay. Most of these sets are ok starter sets and contain slides with assorted plants, insect parts and animal hair. A wide array of prepared slides can be bought at eBay and other online outlets such as science stores - when browsing the goods you are likely to encounter the following abbreviations: 'st' means stained (samples are stained with various chemicals to make it easier to observe certain parts of the sample, as an example the Gram stain stains some bacteria), 'wm' means whole mount (you get the complete organism on your slide), 'ls' is a longitudinal section (section cut lengthwise), 'cs' is a cross section, 'sq' is a squash preparation, and 'sm' is a smear, for example a blood smear.

Thin section mineral slides tend to be somewhat expensive, which makes them less than ideal when you are making your first steps into polarized light microscopy. But worry not, because it is possible to make your own microscope slides for polarized light microscopy at a fraction of the cost of thin section mineral slides.

The key to this is 'optical activity', which is a property of a group of chemical substances called 'chiral molecules'. Optical activity means that the chemical substances can rotate polarized light, just like some minerals can.

Non-chemists might not have heard of chiral molecules before, but nevertheless some of these molecules are part of all of our lives. One of the most widespread is sucrose, also known as white table sugar. Others include glucose (grape sugar), fructose (fruit sugar), tartaric acid (used to add sharpness to some soft drinks) and monosodium glutamate (used as a flavor enhancer, some Asian markets stock it). Your may not be able to grab all of these at your local store, but a bit of table sugar can get you started, and will let you make slides which will look very beautiful when viewed using your polarized light microscope.

Making slides for polarized light microscopy

Dissolve 2 grams (or half a teaspoon if you don't have a scale) of table sugar in 20 grams water (4 teaspoons). Transfer three to four drops to a clean microscope slide, spread the liquid over the slide and put it aside until it has dried completely. Do not use a cover slip. Your slide should now look somewhat like this one:

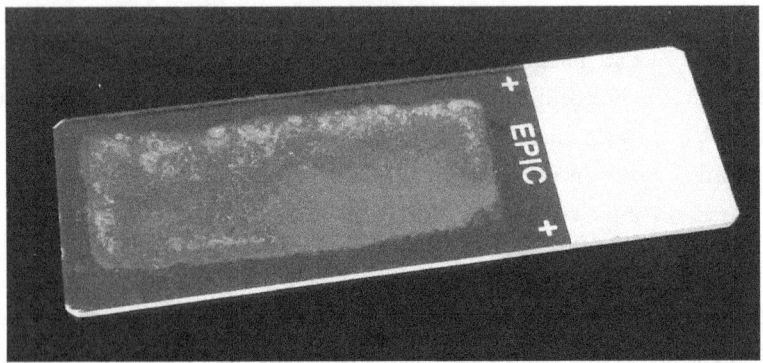

You can repeat this for glucose, fructose, tartaric acid and monosodium glutamate, if they are available where you live. Or you can make several with sucrose, every single slide will be unique! Once you've successfully made a slide and examined it with your microscope you can experiment with using more or less sugar in the same amount of water, or by letting your slide dry slower (find a cold spot in the house) or faster (if you are using glass slides this can be accomplished by placing the slide in an oven at 80C/175F for an hour or two, and then let it cool off).

Sucrose crystal in polarized light

x250 magnification

Sucrose crystal in polarized light

x250 magnification

Tartaric acid crystal in polarized light

x250 magnification

Monosodium glutamate crystal in polarized light

x250 magnification

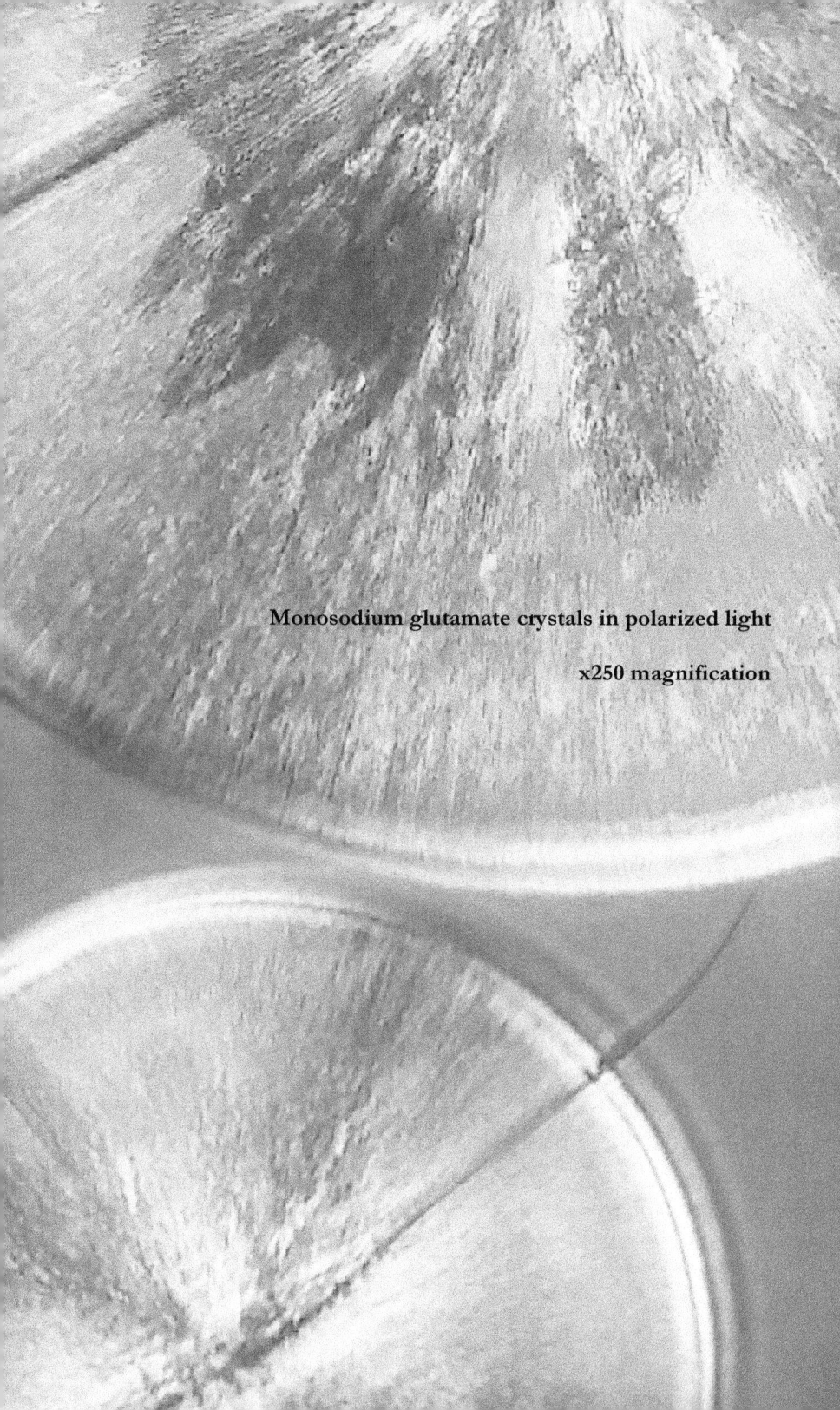

Monosodium glutamate crystals in polarized light

x250 magnification

Additional resources will be available at

alifeinscience.com later in 2016